AF772842

Peter J. Brockwell   Richard A. Davis

ITSM: An Interactive Time Series Modelling Package for the *SPARC* Workstation

With 1 diskette

Springer-Verlag
New York Berlin Heidelberg London Paris
Tokyo Hong Kong Barcelona Budapest

Peter J. Brockwell
Richard A. Davis
Department of Statistics
Colorado State University
Fort Collins, CO 80523
USA

© 1992 Springer-Verlag New York, Inc.
All rights reserved. This work may not be translated or copied in whole or in part without the written permission of the publisher (Springer-Verlag New York, Inc., 175 Fifth Avenue, New York, NY 10010, USA) except for brief excerpts in connection with reviews or scholarly analysis. Use in connection with any form of information storage and retrieval, electronic adaptation, computer software, or by similar or dissimilar methodology now known or hereafter developed is forbidden.
The use of general descriptive names, trade names, trademarks, etc., in this publication, even if the former are not especially identified, is not to be taken as a sign that such names, as understood by the Trade Marks and Merchandise Marks Act, may accordingly be used freely by anyone.

Production and manufacturing managed by Rhea Talbert.
Camera ready copy provided by the authors.
Printed and bound by Intercounty Lithograph Corp., Amityville, NY.
Printed in the United States of America.

9 8 7 6 5 4 3 2 1

ISBN 0-387-14210-X Springer-Verlag New York Berlin Heidelberg
ISBN 3-540-14210-X Springer-Verlag Berlin Heidelberg New York

ITSM 4.0 for the SPARC Workstation

1 Introduction

1.1 The package

The package *ITSM* (Interactive Time Series Modelling) for the SPARC workstation is designed to accompany the book *Time Series: Theory and Methods* by Peter Brockwell and Richard Davis (Springer-Verlag, Second Edition, 1991), referred to below as *BD* . It consists of seven modules whose functions are summarized below.

PEST is a program for the modelling, analysis and forecasting of univariate time series. The name "PEST" is an abbreviation for **P**arameter **EST**imation.

SPEC is a program which performs non-parametric spectral estimation for both univariate and bivariate time series.

SMOOTH permits the user to apply symmetric moving average or one-sided exponential smoothing operators to a given data set. The current version of the program also permits smoothing by removal of high frequency components from the Fourier transform of the data.

TRANS allows the calculation and plotting of sample cross-correlations between two series of equal lengths, and the fitting of a transfer function model to represent the relation between them.

ARVEC fits vector autoregressive models to multivariate time series with up to 6 components and allows automatic order-selection using the AICC criterion.

ARAR is based on the ARARMA forecasting technique of Newton and Parzen. For a univariate data set it first selects and applies (if necessary) a memory-shortening transformation to the data. It then fits a subset autoregressive model to the memory-shortened series and uses the fitted model to calculate forecasts.

LONGMEM can be used to simulate data from a specified fractionally integrated ARMA model with zero mean (*BD* Sect.13.2). It can also be used to fit such a model to a data set (by maximizing the Whittle approximation to the Gaussian likelihood) and to forecast future values of the series.

With a few exceptions, due mainly to the differences between DOS and UNIX, the PC and SPARC versions of *ITSM* are run in the same way. The accompanying manual, *ITSM: An Interactive Time Series Modelling Package for the PC* by Peter Brockwell and Richard Davis, (Springer-Verlag, 1991), can therefore be used with both versions. For information regarding the features specific to the SPARC version, see the example in Section 3 below and Appendix A. For a more extensive discussion of time series modelling and the methods used in *ITSM*, see the book *Time Series: Theory and Methods.*

1.2 System requirements

- The program *ITSM* Version 4.0 will operate on Sun–4 workstations with SunOS release 4.1.1 or above and OpenWindows2. *ITSM* graphics may be viewed at the console of the SPARC or on an X-terminal provided an X11 based window manager (such as OpenWindows) has been initiated. *ITSM* graphics are not available under SunView.
- Graphics output can be saved in a file written in HP PCL and printed on an HP LaserJet II, III or compatible laser printer.

2 Installation and testing of ITSM

2.1 Installation from a diskette

- Log into your account.
- Change the current directory to the one in which you would like to install *ITSM*. Normally this would be either your home directory or a more central one such as */usr/local*. For the latter choice, you will first need to become superuser. The commands required are,
  - for installation in your home directory:

    ```
    mysun% cd ~
    ```

  - for installation in */usr/local* (assuming you are superuser):

    ```
    mysun# cd /usr/local
    ```

- Insert the $3\frac{1}{2}$" diskette into the floppy drive.
- Extract *ITSM* using the tar command:

  ```
  mysun% bar xvfp /dev/rfd0
  ```

- This step should have created a subdirectory *itsm4* in the current directory. Now change to the new subdirectory *itsm4*:

  ```
  mysun% cd itsm4
  ```

- Run the program *setup*:

  ```
  mysun% setup
  ```

  The following message, with *install-dir* replaced by the name of the directory where *ITSM* was installed, should appear on the screen.

  ```
  ITSM should be installed in the directory install-dir/itsm4
  If this is not the case, rerun setup from the itsm4 subdirectory.
  ```

  Check to see that a file called *itsm* containing the 3 lines

  ```
  ITSM=install-dir/itsm4
  export ITSM
  install-dir/itsm4/xitsm
  ```

was created (issue the command `cat itsm` to view the contents of *itsm*). If for some reason the file *itsm* was not created properly or not created at all, copy the backup version of *itsm*, called *itsm.back* into *itsm* using the command `cp itsm.back itsm`. Then read the file *itsm* into an editor and follow the instructions given in the file.

- Remove the diskette:

```
mysun% eject
```

- You can now execute *ITSM* from the subdirectory *~/itsm4* or */usr/local/itsm4*, depending on where it was installed. To execute *ITSM* from any directory you can either

  - copy the file *itsm* in the *itsm4* subdirectory to a directory which is already in your path. (To determine your current path, type `echo $PATH`.) For example, either of the following 2 commands (which can only be issued by a superuser) should do the trick.

    ```
    mysun# cp itsm /usr/local
    mysun# cp itsm /usr/bin
    ```

    Both these options will allow access to *ITSM* for all users even if *ITSM* was not installed in */usr/local*.

  or

  - add the new *itsm4* directory to your path specification.

2.2 Testing from the console

To test that *ITSM* was properly installed, log out and log back in making sure that an X–based windowmanager (such as OpenWindows) has been started. Assuming that the file *itsm* is located in a directory contained in your path, initiate an *ITSM* session from a work window as follows:

```
mysun% itsm
```

The following message should appear on your screen.

```
                    I T S M
Copyright (C)  1992  B & D Enterprises, Inc.
(Version 4.0 Release 1 for Sun4, Mar 1992)





ITSM graphics are only available on a terminal
running X-Windows.

Do you want graphics for this session (y/n)?
```

Respond to the question by entering **y** (i.e. by typing **y** then pressing the return key). If *ITSM* has been installed properly, then two new windows, *ITSM Graph Window 1* and *ITSM Graph Window*

2, will be created and the *Module Menu* with its nine options will appear in the window from which *ITSM* was executed.

ITSM will run without graphics capability on any remote terminal provided you answer no to the question, `Do you want graphics for this session (y/n)?`

2.3 Testing on an X-terminal or a remote machine

For many configurations of X-terminals and networks, *ITSM* 4.0 graphics can be displayed across the network on an X-terminal. Begin by starting the window manager on the X-terminal and open a work window. Before you log into (*mysun*) on which *ITSM* was installed, *mysun* needs to have access to the display on your X-terminal. To make sure this is the case use the command

```
xhost mysun
```

(See your systems manager if you are unable to issue this command.) Now log into *mysun* (`rlogin mysun` or `telnet mysun`). The computer *mysun* may also need to know the name of your X-terminal. If this is the case or if you are unsure, use the command

```
mysun% setenv DISPLAY xterminal:0
```

where *xterminal* is the IP number or name of the X-terminal. (On some networks, it is not necessary to issue either the `xhost` or `setenv` commands.) To start an *ITSM* session type

```
mysun% itsm
```

and continue as described in the previous section.

3 An Example

In this example, we illustrate the following points using the module *PEST* operated from the console of the machine containing *ITSM* :

- reading a data file from the *itsm*4 directory into *PEST*.
- graphing a data set.
- printing a graph on an HP LaserJet printer.

Begin by logging into your account and starting an X-based window manager. Open a work window and initiate an *ITSM* session by typing `itsm` (this assumes that the file *itsm* is located in a directory contained in your path or that the current directory is the *itsm*4 subdirectory). After answering yes (type `y` followed by a carriage return) to the question related to graphics display, *ITSM Graph Window 1* and *ITSM Graph Window 2* will appear on the screen. It is advantageous to reposition the three windows so that the work window is on the right side of the screen and the two graphics windows occupy the left side of the screen one above the other. The Module Menu should now be displayed in the work window. Choose Option 1: PEST from the Module Menu by typing `1` followed by a carriage return. You will now see the Main Menu of the *PEST Module.* Select Option 1 (Data entry; statistics; transformations) and the Data Menu will then appear. Again chose Option 1 and you will be asked to confirm that you wish to enter new data. Respond by entering **y** and you will be asked for the name of the new data file. To read in the *airpass* data located in the *itsm*4 directory type

```
$ITSM/airpass
```

(Pay attention to the $ and the capitalization since unlike DOS, UNIX makes a distinction between upper and lower case letters.) All of the data files in the *itsm*4 directory (see Appendix B) can be accessed in the *ITSM* modules using the file designation **$ITSM/filename**.

To plot the data $ITSM/airpass in *ITSM Graph Window 1*, choose Option 2 of the Data Menu. Unless you want to alter the graph by rescaling or plotting a subset, enter 0. To print the graph click on the print box in the upper right corner of the graph window. A small rectangular window will open in the middle of the graph window requesting the filename under which to store the graph. Enter the file name **airpass.hp** followed by carriage return (the pointer must be in the subwindow while entering the filename). Assuming you have an HP LaserJet printer attached to your workstation or network, the file *airpass.hp* can either be printed now or after you have completed the *ITSM* session. To print the file now, temporarily suspend the *ITSM* session by holding down the **<Control>** key and pressing the **Z** key (the suspend sequence may vary from machine to machine). At the prompt, issue the command

```
mysun% lpr -Php -r -l airpass.hp
```

where *hp* is the name of the HP LaserJet printer (see the systems administrator to find the correct specification for the HP LaserJet on your system), the *−r* flag removes the file once it has been printed, and the *−l* flag is needed for transmission of a binary file. If you would like to keep the file, delete the *−r* flag from the command line. If the printer is set to receive graphics files written in the HP printer language pcl, you may also be able to use the print command

```
mysun% lpr -Php -r airpass.hp
```

(assuming the *airpass.hp* file has not yet been removed). If neither of these options work contact your systems administrator. To return to the *ITSM* session right where you left it, enter **fg** for *foreground.* Since you left *ITSM* waiting for your selection from the Data Menu, choose an item from the menu to continue. If you decide to print the graphics files at the conclusion of an *ITSM* session, you simply issue the same command as above for each of the files that were created.

The default setting for *ITSM* allows for graphs to be displayed alternately in the two graphics windows. This feature may be deactivated giving you the option of directing each graph to a chosen window. To deactive this option select Option 8: Change window options from the Module Menu and change window allocation to MANUAL.

Consult Chapters 2–7 of the manual, *ITSM: An Interactive Time Series Modelling Package for the PC*, for more detailed information on the use of *ITSM*.

Appendix A

In this appendix, differences between ITSM for the SPARC workstation and ITSM Version 3.0 for the IBM PC are indicated.

ARVEC

1. The module ARVEC has been upgraded by the inclusion of Jones's multivariate version of the Burg algorithm as an alternative to autoregressive model-fitting via the Yule-Walker equations and Whittle's algorithm.

   The calculation and minimization of the multivariate AICC for order selection have been corrected in the current version of ARVEC. More details are given in the following paragraphs which describe the program.

   The steps required to run ARVEC are quite analogous to those described in Section 3 above for running PEST. The program will read n observations of an m-variate time series (where $n \leq 1000$ and $m \leq 6$), allow up to two differencing operations, and then subtract the component means to give the series,

   $$\mathbf{X}_t = (X_{t,1}, X_{t,2}, \ldots, X_{t,m})', t = 1, 2, \ldots.$$

   It next fits a multivariate AR process of specified order $p < 21$ to the multivariate series $\mathbf{X}$ and predicts future values, giving estimated mean squared errors for the predicted values. The fitted AR model is

   $$\mathbf{X}_t = \mathbf{\Phi}_1 \mathbf{X}_{t-1} + \cdots + \mathbf{\Phi}_\mathbf{p} \mathbf{X}_{\mathbf{t-p}} + \mathbf{Z}_\mathbf{t},$$

   where $\mathbf{\Phi}(j), j = 1, \ldots, p,$ are $m \times m$ matrices and $\{Z(t)\}$ is multivariate WN(0,V).

   There is a choice of two estimation algorithms. The first is Jones's multivariate version of the Burg algorithm (*Applied Time Series Analysis*, ed. Findley, Academic Press, 1978), and the second is Whittle's multivariate version of the Durbin-Levinson algorithm (*BD* ,Section 11.4). The order p can be specified or the program can be used to find the value of p which minimizes the AICC statistic.

   For the Burg model the AICC is computed as

   $$n \ln(\det(V)) + nm(1 + ln(2\pi)) + (2pm^2 + 2)/(1 - (pm^2 + 2)/nm),$$

   and for the Yule-Walker model it is computed as

   $$-2\ln(L) + (2pm^2 + 2)/(1 - (pm^2 + 2)/nm),$$

   where L is the Gaussian likelihood.

   If differencing was performed, the program will also predict the original series and find estimated mean squared errors of the predictors.

PEST

1. In *ITSM* Version 3.0 for the IBM PC, the preliminary ARMA models fitted in Option 3 of PEST are models for the *mean-corrected* data, even though the data currently stored in PEST may not have been mean-corrected. In the SPARC version of the program this has been changed. Option 3 now assumes that the current data are from a series with mean equal to zero (the sample covariances used for preliminary estimation are now $\tilde{\gamma}(h) = n^{-1}\sum_{r=1}^{n-|h|} x_r x_{r+|h|}$ instead of $\hat{\gamma}(h) = n^{-1}\sum_{r=1}^{n-|h|}(x_r - \overline{x})(x_{r+|h|} - \overline{x})$ where $\{x_t\}$ is the *current* data set). This makes Option 3 consistent with Option 8 in the sense that they both fit (zero-mean) ARMA models to the data currently in PEST. If you wish to use either option to fit an ARMA model to a mean-corrected series, you must first subtract the mean from the data in Option 1 (**Data entry; statistics; transformations**) and then go to Option 3 or Option 8.

   Example

   For the SUNSPOT series $\{x_t\}$, if we go to Option 3 of PEST without first subtracting the sample mean (46.93) from the data, and fit a preliminary AR(2) model using the Yule-Walker Option, we obtain the model,

$$X_t - 1.284X_{t-1} + 0.413X_{t-2} = Z_t, \quad \{Z_t\} \sim WN(0, 336.9),$$

   for $\{x_t\}$. This is a poor model however since the mean of the SUNSPOT series is clearly non-zero. We should therefore subtract the mean from the data using Option 1. The current data set in PEST then becomes $\{y_t = x_t - 46.93\}$. Returning to Option 3 and again using the Yule-Walker method, we obtain the model,

$$Y_t - 1.318Y_{t-1} + 0.634Y_{t-2} = W_t, \quad \{W_t\} \sim WN(0, 232.9),$$

   for the mean-corrected series $\{y_t\}$, which is also the data set currently stored in PEST. (ITSM Version 3.0 for the IBM PC gives the preliminary estimates 1.318, -0.634 and 232.9, regardless of whether $\{x_t\}$ or $\{y_t\}$ is the current data set.)

2. Burg's algorithm for fitting univariate AR(p) models has been incorporated as an alternative to Yule-Walker estimation in PEST (Option 3).

3. As in *ITSM* Version 3.0 for the IBM PC, the estimated covariance matrix of the maximum likelihood estimators found in Option 8 of PEST is found by numerical computation of the Hessian matrix at the maximum (see *BD* , Section 9.2). If the numerically evaluated Hessian matrix turns out to be not positive definite and the model is invertible, the program will now compute the asymptotic approximation to the covariance matrix found in *BD* , Section 8.8. The SPARC version prints the estimated correlation matrix of the coefficient estimators instead of the corresponding covariance matrix. (Both versions also print the estimated standard deviations.)

SMOOTH

1. An additional procedure for data-smoothing has been included in SMOOTH as Option 3 of the Main Menu. It computes the discrete Fourier transform,

$$a_j = n^{-1/2}\sum_{t=1}^{n} x_t e^{-it\omega_j}, \quad \omega_j = \frac{2\pi j}{n}, \quad -\frac{n-1}{2} \le j \le \frac{n}{2},$$

of the data $x_1, \ldots, x_n$, and sets to zero the Fourier coefficients corresponding to frequencies greater in absolute value than $f\pi$, where f is a parameter between 0 and 1, specified when you run the program. (For maximum smoothing set $f = 0$.) The resulting transform is then inverted to produce the smoothed data,

$$s_j = n^{-1/2} \sum_{j:|\omega_j| \le f\pi} a_j e^{it\omega_j}, \quad j = 1, \ldots, n.$$

LONGMEM

1. The new module *LONGMEM* is designed for simulation, model-fitting and prediction with ARIMA(p, d, q) processes, where $-.5 < d < .5$. Such processes, known as fractionally integrated ARMA processes, are stationary solutions of difference equations of the form,

$$(1 - B)^d \phi(B) X_t = \theta(B) Z_t, \quad \{Z_t\} \sim WN(0, \sigma^2),$$

where the autoregressive and moving average polynomials, $\phi(z)$ and $\theta(z)$ satisfy

$$\phi(z) \ne 0 \text{ and } \theta(z) \ne 0 \ \forall z \text{ such that } |z| \le 1.$$

(See *BD* Section 13.2 for more details.)

The autocovariance function of the model is computed from the formula (*BD* p.525),

$$\gamma_X(h) = \sum_{j=0}^{\infty} \sum_{k=0}^{\infty} \psi_j \psi_k \gamma_Y(h + j - k),$$

where $\sum_{i=0}^{\infty} \psi_i z^i = \theta(z)/\phi(z)$, $|z| \le 1$, and $\gamma_Y(\cdot)$ is the autocovariance function of fractionally integrated white noise with parameters d and σ^2 (see *BD* , equations (13.2.8) and (13.2.9)). The upper limits of summation in the expression for $\gamma_X(h)$ are replaced by M, where M has the default value of 50. If $\phi(z)$ has a zero close to the unit circle you may wish to increase the value of M to reduce the truncation error. The value of M may be increased to any value less than or equal to 200.

Simulation (of up to 500 observations) makes use of the innovations algorithm as described in *BD* Problem 8.16.

Prediction is also carried out using the innovations algorithm as described in *BD* Section 5.3. Estimation of the parameters $\boldsymbol{\beta} = (d, \phi_1, \ldots, \phi_p, \theta_1, \ldots, \theta_q)'$ and σ^2, is by maximization of the Whittle approximation, L_W, to the Gaussian likelihood, or equivalently by minimization of

$$-2\ln(L_W) = n\ln(2\pi) + 2n\ln\sigma + \sigma^{-2} \sum_j \frac{I_n(\omega_j)}{g(\omega_j; \boldsymbol{\beta})} + \sum_j \ln g(\omega_j; \boldsymbol{\beta}),$$

where I_n is the periodogram, $\sigma^2 g$ is the model spectral density and $\sum_j$ denotes the sum over all non-zero Fourier frequencies, $\omega_j = 2\pi j/n \in (-\pi, \pi]$.

Appendix B

The following data sets are included with the SPARC version of *ITSM* . To access these data files from within *ITSM* , use the filename specification **`$ITSM/filename`** where filename is the name (in *lower* case letters) of one of the data sets listed below.

Data Sets

airpass International airline passenger monthly totals (in thousands), Jan. 49 – Dec. 60. From Box and Jenkins (*Time Series Analysis: Forecasting and Control, 1970*). *BD Example 9.2.2.*

appa Lake level of Lake Huron in feet (reduced by 570), 1875–1972. *BD Appendix Series A.*

appb Dow Jones Utilities Index, Aug.28–Dec.18, 1972. *BD Appendix Series B.*

appc Private Housing Units Started, U.S.A. (monthly). From Makridakis competition, series 922. *BD Appendix Series C.*

appd Industrial Production, Austria (quarterly). From Makridakis competition, Series 337. *BD Appendix Series D.*

appe Industrial Production, Spain (monthly). From Makridakis competition, Series 868. *BD Appendix Series E.*

appf General Index of Industrial Production (monthly). From Makridakis competition, Series 904. *BD Appendix, Series F.*

appg Annual Canadian Lynx Trappings, 1821–1934. *BD Appendix Series G.*

apph Annual Mink Trappings, 1848–1911. *BD Appendix Series H.*

appi Annual Muskrat Trappings, 1848–1911. *BD Appendix Series I.*

appj Simulated input series for transfer function model. *BD Appendix Series J.*

appjk.2 The two series APPJ and APPK (see above) in bivariate format for analysis by ARVEC.

appk Simulated output series for transfer function model. *BD Appendix Series K.*

beer Monthly beer production in Australia, (in megalitres), Jan. 56 – June 88.

chocs Monthly confectionery production in Australia, (in tons), July 57–June 88.

deaths Monthly accidental deaths in the U.S.A., 1973–1978 (National Safety Council). *BD Example 1.1.6.*

e911 200 simulated values of an ARIMA(1,1,0) process. *BD Example 9.1.1.*

e921 200 simulated values of an AR(2) process. *BD Example 9.2.1.*

e923 200 simulated values of an ARMA(2,1) process. *BD Example 9.2.3.*

e951 200 simulated values of an ARIMA(1,2,1) process. *BD Example 9.5.1.*

e1021 Sinusoid plus simulated Gaussian white noise. *BD Example 10.2.1.*

e1042 160 simulated values of an MA(1) process. *BD Example 10.4.2.*

e1321 200 values of a simulated fractionally integrated MA(1) series. *BD Example 13.2.1.*

e1331 200 values of a simulated MA(1) series with standard Cauchy white noise. *BD Example 13.3.1.*

e1332 200 values of a simulated AR(1) series with standard Cauchy white noise. *BD Example 13.3.2.*

elec Monthly electricity production in Australia (in KWH), Jan. 56–June 88.

finserv Quarterly Australian expenditures on financial services (in millions of dollars), Sept. qtr 69 – June qtr 88.

imports Australian imports from Japan (in thousands of dollars), July 65 – June 88.

lead Leading Indicator Series from Box and Jenkins (*Time Series Analysis: Forecasting and Control, 1970*). *BD Example 11.2.2.*

lres Whitened Leading Indicator Series obtained by fitting an MA(1) to the mean-corrected and differenced LEAD series and storing the residuals. *BD Section 13.1.*

ls.2 Lead-Sales data in bivariate format for analysis by ARVEC.

sales Sales Data from Box and Jenkins (*Time Series Analysis: Forecasting and Control, 1970*). *BD Example 11.2.2.*

sres Residuals obtained from the mean-corrected differenced SALES data when the filter used for whitening the mean-corrected differenced LEAD series is applied. *BD Section 13.1.*

strikes Strikes in the U.S.A., 1951–1980 (Bureau of Labor Statistics). *BD Example 1.1.3.*

sunspots The Wolfer sunspot numbers, 1770–1869. *BD Example 1.1.5.*

uspop U.S. Population at ten-year intervals, 1790–1980 (U.S.Bureau of the Census). *BD Example 1.1.2.*